101 Applications of Crystal Field Theory

Christoph Sontag PhD

Copyright © 2018 Christoph Sontag

All rights reserved.

ISBN: 1544857403
ISBN-13: 978-1544857404

DEDICATION

To write this booklet would not have been possible without the help of my wife Uan. I dedicate this work to her.

CONTENTS

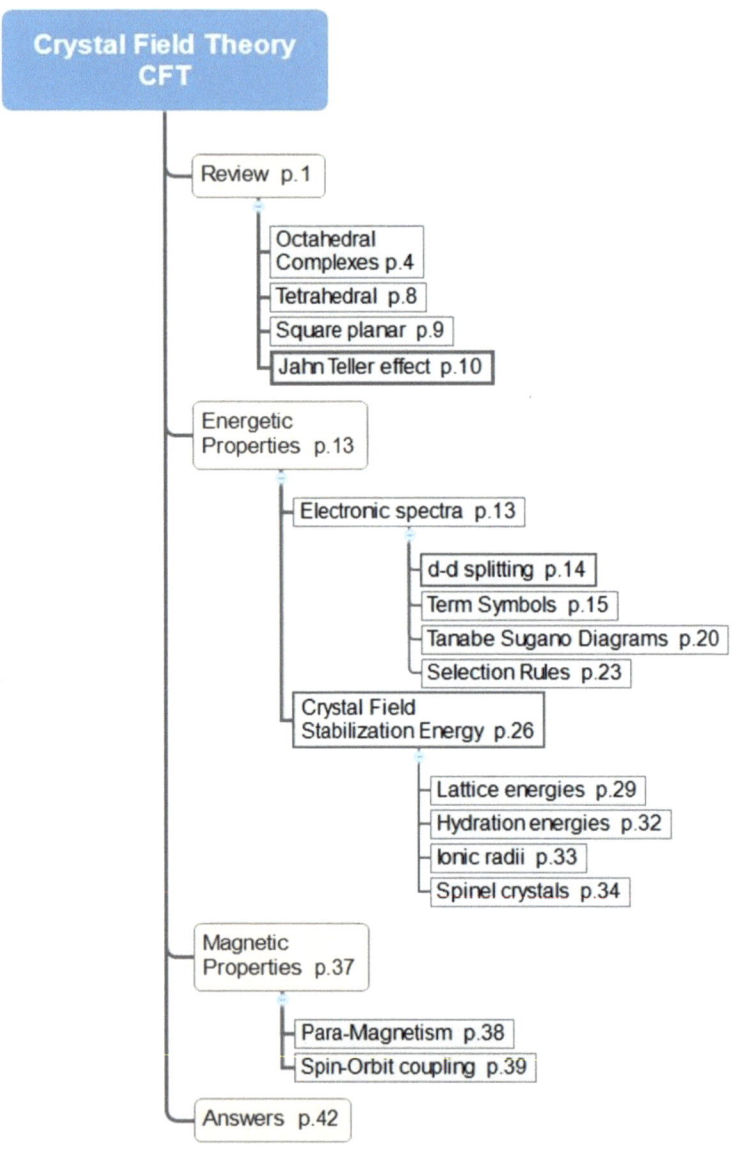

1 INTRODUCTION: WHY "CRYSTAL" ?

The "crystal field theory" is a simple bonding model for transition metal complexes, that focuses on the interaction between a metal ion d-orbitals (like Cu^{2+}) and ligands (normally with an electron lone pair) like H_2O or NH_3.

As an example we can look at copper sulfate: this salt comes with a blue color, but when we heat it, it loses the crystal water and it turns white.

The absorbed so-called "crystal water" is built into the crystal structure of the molecule, obviously changing its electronic structure.

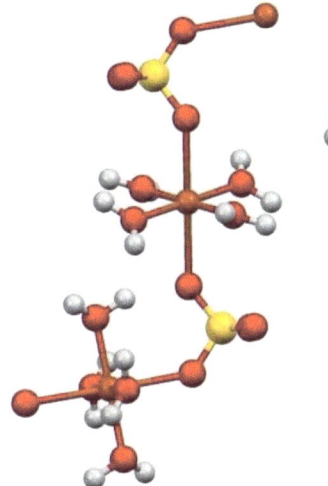

The structure* shows that the copper ions are surrounded by six oxygen atoms (two from sulphate, four from water) forming an octahedral complex.

This environment causes the effect that **red** color is absorbed by the solid, let it appear **blue**.

* *By Smokefoot - Own work, CC BY-SA 4.0,*
https://commons.wikimedia.org/w/index.php?curid=53778827

In other words, the surrounding oxygen atoms create an energy gap in the orbitals of the copper ion. We should remember that visible color is produced when an object absorbs a part of the visible light spectrum:

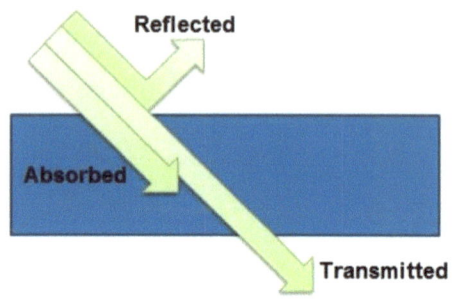

A surface appears to have a color because it absorbs part of the visible light

(*http://study.com/academy/lesson/the-absorption-coefficient-definition-calculation.html*)

For example, the surface of a leaf appears green because the chlorophyll absorbs the complementary color (red) from the whole spectrum of the white light.
(*http://www.imagewithjoy.com/external-self/coloring/color-theory-basics*)

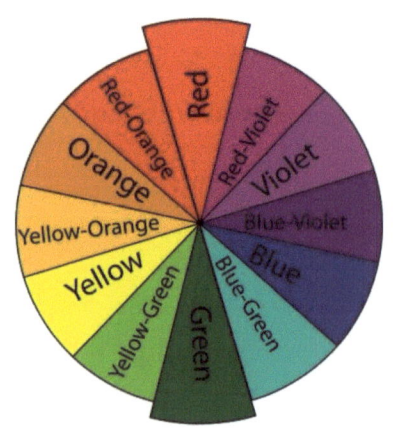

Color	Wavelength
violet	380–450 nm
blue	450–495 nm
green	495–570 nm
yellow	570–590 nm
orange	590–620 nm
red	620–750 nm

Approximate wavelengths of each color

(*https://en.wikipedia.org/wiki/Visible_spectrum*)

How is light absorbed ?

The energy of certain wavelengths is absorbed by electronic changes in the material, moving electrons from a ground into an excited state.

One example is the energy absorption of a non-bonding electron in a molecule which then moves to an anti-bonding orbital.

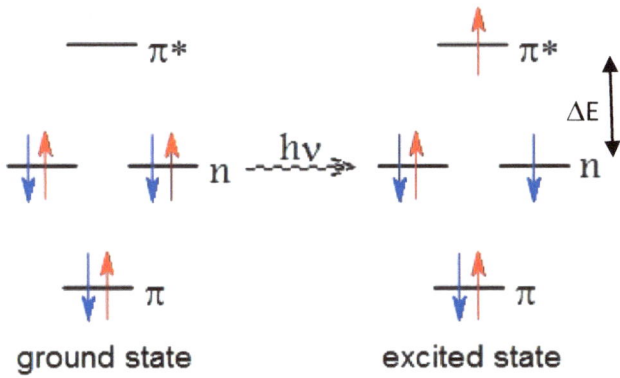

ground state   excited state

http://people.chem.ucsb.edu/kahn/kalju/chem126/public/elspect_theory.html

The energy difference ΔE between the n- and the π^* orbital corresponds to the energy of the light which is absorbed:

$\Delta E = h\nu = h * c/\lambda$

(h: Planck's constant, ν frequency, c speed of light = $3*10^8$ m/s, and λ wavelength)

For spectroscopic calculations where we deal with single photons, it is convenient not to follow the SI units, but instead use electron Volt as energy unit. This is the energy that one electron gains when it is accelerated in an electric field of 1 volt.

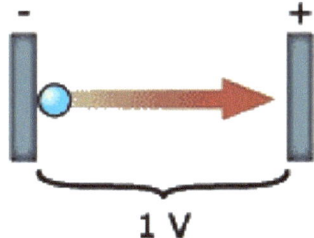

[1] *Calculate the energy gap in eV and in cm^{-1} of the blue $CuSO_4*5H_2O$ ($h = 4.14 *10^{-15}$ eV·s, wavenumber $\tilde{\nu} = 1/\lambda$ in cm^{-1})*

[2] *Explain the color of chlorophyll b and calculate the energies that are absorbed in the visible spectrum in eV.*

(http://hyperphysics.phyastr.gsu.edu/hbase/Biology/ligabs.html)

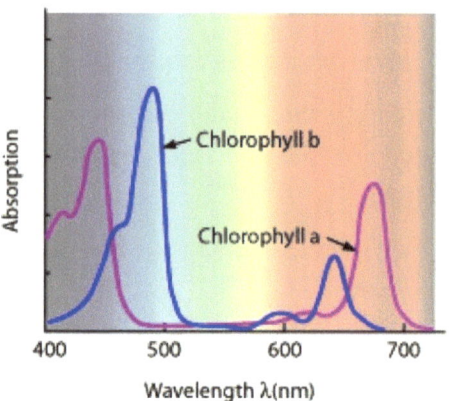

2 REVIEW OF CRYSTAL FIELD THEORY
2.1 OCTAHEDRAL COMPLEXES

The crystal field theory (CFT) can explain this energy gap resulting from "splitting" of the d-orbitals of the metal ion by the approaching negative charges of the ligand lone pairs.

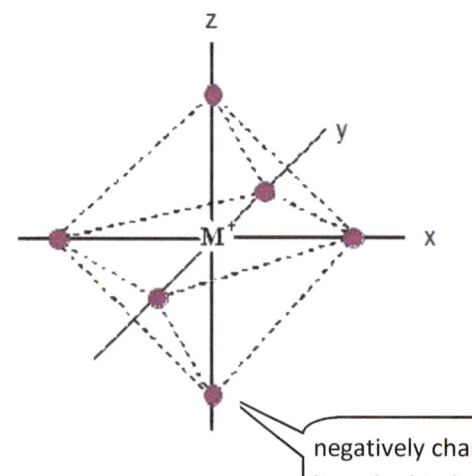

[3] *Draw the $d_{x^2-y^2}$ and d_{xy} orbitals around the metal center – in which case is a closer contact between the metal and the ligand orbitals?*

Compare also the d_{z^2} and the d_{xz} orbital.

negatively charged ligand orbitals

You may have discovered that the two d-orbitals with maxima on the axis (d_{x2-y2} and d_{z2}) have a close contact to the ligands:

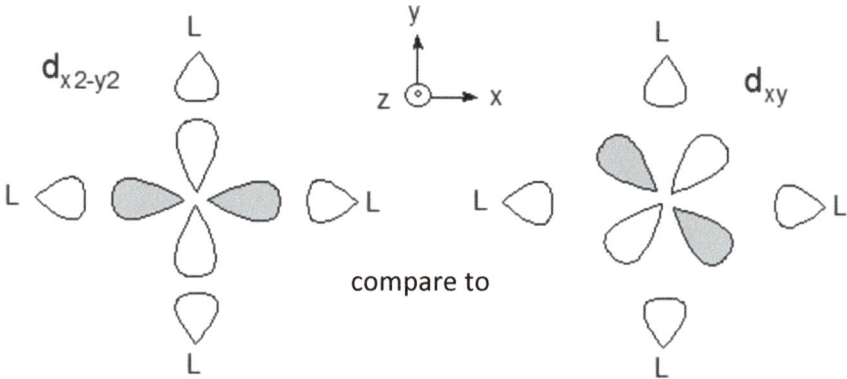

compare to

The other three d-orbitals (d_{xy}, d_{xz} and d_{yz}) have less contact.

A close contact between the electron densities of the ligand lone pairs and the metal d-orbitals causes a <u>repulsion</u> – or in other words causes a higher energy of these d-orbitals than a less close contact.

Therefore the three orbitals in the "middle" of the ligands (xy, xz and yz) have a lower energy than the two orbitals on the axis (x^2-y^2 and z^2):

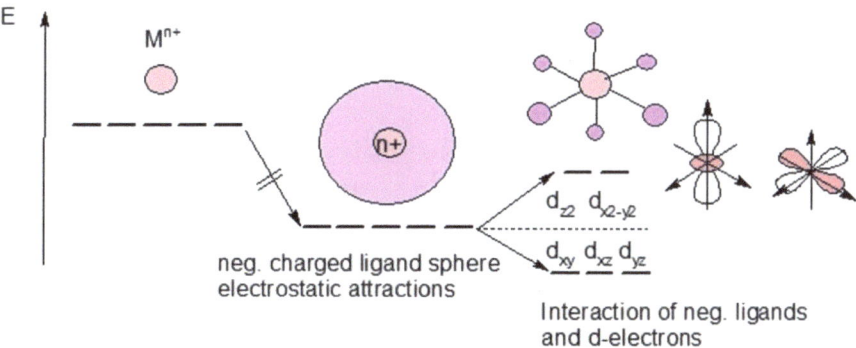

Splitting the metal d-orbitals results in an energy gap called Δo ("o for octahedral) that makes it possible for electrons to absorb light energy by going up to the higher level.

Checking out the symmetries of the d-orbitals in an octahedral symmetry, we can consult the character table *:

* see: "101 Group Theory" of this series of tutorials

Character table for O_h point group

	E	$8C_3$	$6C_2$	$6C_4$	$3C_2 = (C_4)^2$	i	$6S_4$	$8S_6$	$3\sigma_h$	$6\sigma_d$	linear, rotations	quadratic
A_{1g}	1	1	1	1	1	1	1	1	1	1		$x^2+y^2+z^2$
A_{2g}	1	1	-1	-1	1	1	-1	1	1	-1		
E_g	2	-1	0	0	2	2	0	-1	2	0		$(2z^2-x^2-y^2, x^2-y^2)$
T_{1g}	3	0	-1	1	-1	3	1	0	-1	-1	(R_x, R_y, R_z)	
T_{2g}	3	0	1	-1	-1	3	-1	0	-1	1		(xz, yz, xy)
A_{1u}	1	1	1	1	1	-1	-1	-1	-1	-1		
A_{2u}	1	1	-1	-1	1	-1	1	-1	-1	1		
E_u	2	-1	0	0	2	-2	0	1	-2	0		
T_{1u}	3	0	-1	1	-1	-3	-1	0	1	1	(x, y, z)	
T_{2u}	3	0	1	-1	-1	-3	1	0	1	-1		

[4] assign the symmetries of all five d-orbitals

We can see that the two groups of d-orbitals are characterized by two different symmetries – e_g and t_{2g}.

This splitting of d-orbitals is quite small, so colored light is enough to excite electrons to the higher e_g level:

This of course happens only if there is some space on this level, therefore Zn^{2+} ions with 10 d-electrons cannot have a color.

In the case of a d^1 complex like Ti^{3+} in water, we can expect a single peak in the UV/VIS spectrum:

The absorption maximum wavelength corresponds to the energy gap Δo.

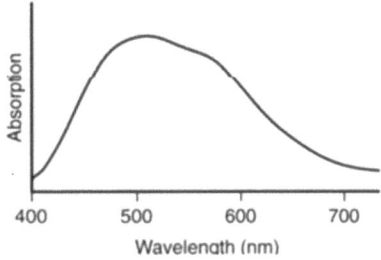

(http://chemistry.stackexchange.com/questions/53252/clarification-on-absorption-spectra-and-crystal-field-theory)

[5] *Calculate Δo in cm^{-1} for this Ti^{3+} complex*

(you may observe that there is a shoulder peak at higher wavelength – which belongs to another complex configuration caused by Jahn-Teller distortion – see chapter2.4)

2.2 TETRAHEDRAL COMPLEXES

Early and late transition metals (Sc/Ti and Co/Ni groups) prefer a four coordination ML$_4$, which can be either tetrahedral or square planar. First row d-metals mostly form tetrahedral complexes because the metal ion is small and the tetrahedral configuration allows more space for the ligands:

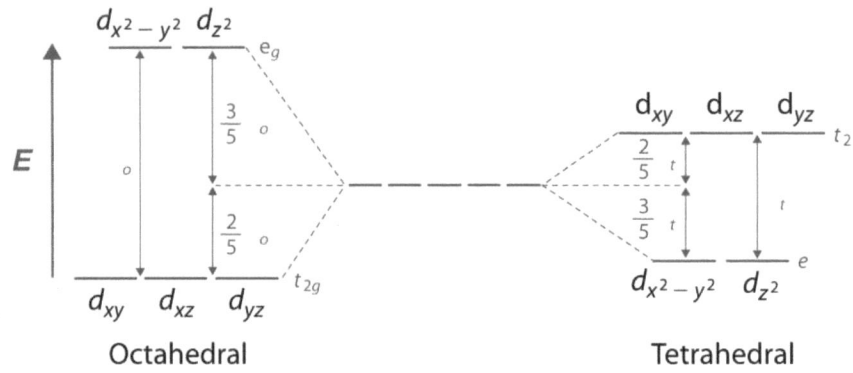

[6] *Draw the d_{z^2} and the d_{xz} orbitals – which one comes in closer contact to the ligands ?*

This is harder to see than for the octahedral configuration – we do not have a direct close contact but we can perceive that now the group of three d-orbitals xy, xz and yz are closer to the ligands than the other group of x^2-y^2 and z^2.

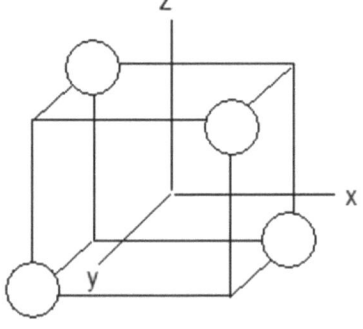

Therefore the energies of the three orbitals between the axis (xy, xz and yz) are now higher than the two orbitals on the axis (x^2-y^2 and z^2).

Because all the d-orbitals do not come in a close contact to the ligands, the energy splitting in tetrahedral configuration is only 0.44 of the octahedral.

[7] Explain why tetrahedral complexes are always high-spin.

2.3 SQUARE PLANAR COMPLEXES

The tetrahedral configuration has the advantage that the ligands have maximum distance to each other, and therefore is preferred for small center metal ions like Co and Ni. But when the ligands can come into the same plane, the molecule can profit from a further crystal field stabilization energy. We can derive the d-orbital energies from the octahedral configuration by removing the upper and lower ligand. In this way, the interaction between d_{z2} and ligands is minimized, instead between d_{x2-y2} and the four ligands in the plane becomes strong:

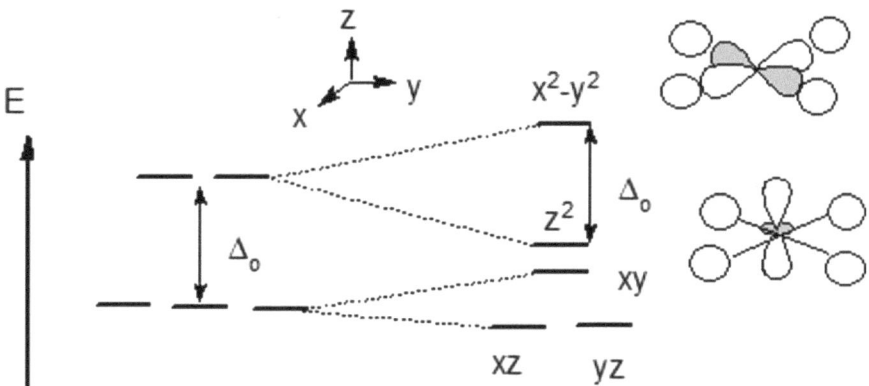

For the remaining 3 d-orbitals there is only a little change – the orbitals with a z-component get lowered in energy and the xy orbital raises slightly.

[8] From the energy diagram, which electron count is most preferred in

an square-planar configuration compared to tetrahedral ?

This configuration is typical for metals of the Co and Ni group with 16 valence electrons.

2.4 GEOMETRIC DISTORTION (JAHN-TELLER EFFECT)

Dependent on the electron configuration, it can be preferable for an octahedral complex to get distorted and change its energy diagram towards the square-planar form:

By elongation of the axial bonds, the electron can gain some more CFSE. This can be seen in the spectrum:

VIS spectrum of $TiCl_3$ in water

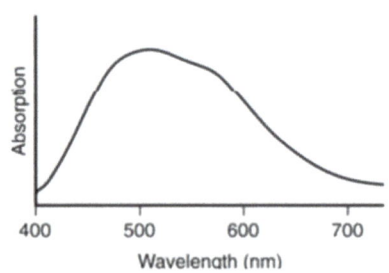

Instead of a sharp peak at 520 nm we observe two peaks close together. These can be attributed to two transitions:

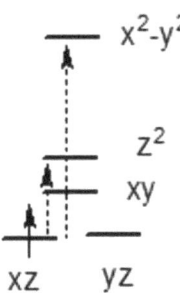

The same effect happens by elongation of the 4 bonds in the plane:

In the same way, many complexes undergo this kind of deformation to lower their overall CFSE.

[9] *Explain why Cu(II) complexes have extra stabilization compared to Ni(II) and Zn(II):*

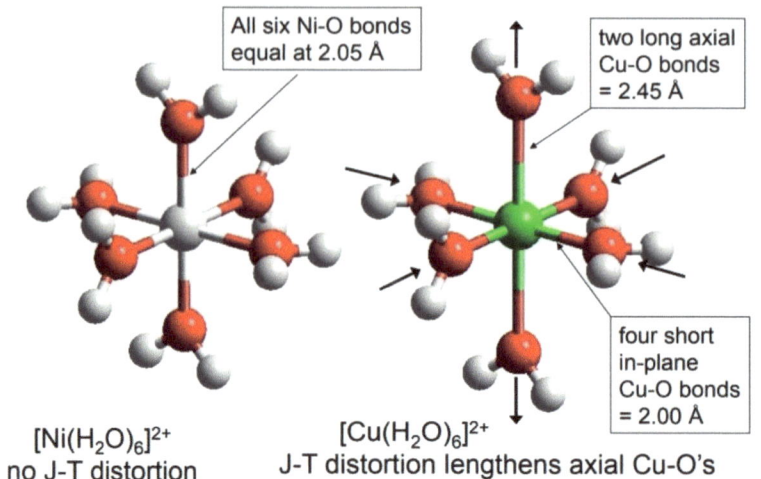

[Ni(H$_2$O)$_6$]$^{2+}$
no J-T distortion

[Cu(H$_2$O)$_6$]$^{2+}$
J-T distortion lengthens axial Cu-O's

(http://www.uniroma2.it/didattica/ChimInorg/deposito/lezione29.pdf)

[10] *Explain why a high-spin d^4 octahedral complex is geometry distorted in contrast to a low-spin d^4*

3 COLORS – ELECTRONIC SPECTRA

Typically the energies splitting for d-orbitals is quite small, so that the energy of visible light is normally enough to cause an electron moving up to a higher level, thereby absorbing light energy:

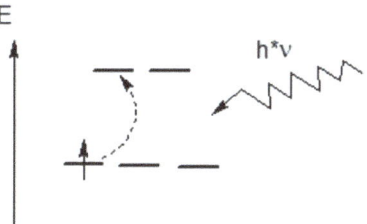

This means that the visible color of the molecule is the <u>composite</u> color that is absorbed by it:

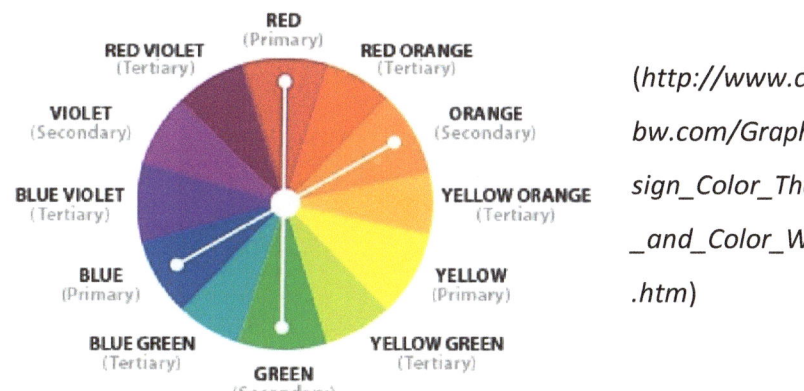

(http://www.co-bw.com/GraphicDesign_Color_Theory_and_Color_Wheel.htm)

So for example, when the light absorbed has a high energy corresponding to blue color, then the complex will look orange.

[11] *When the energy of the d-d splitting increases, does the color of the complex have a blue- or a red-shift ?*

3.1 RELATION BETWEEN D-D SPLITTING AND Δo

In the above case of a d^1 complex, the absorption maximum in the VIS-spectrum corresponds directly to the splitting energy Δo.

But when there are several electrons, the situation becomes more complicated since then we have to deal with electron-electron interaction as well. In case of a d^2 complex like V^{3+} there are **two** absorption maxima in the spectrum instead of only one:

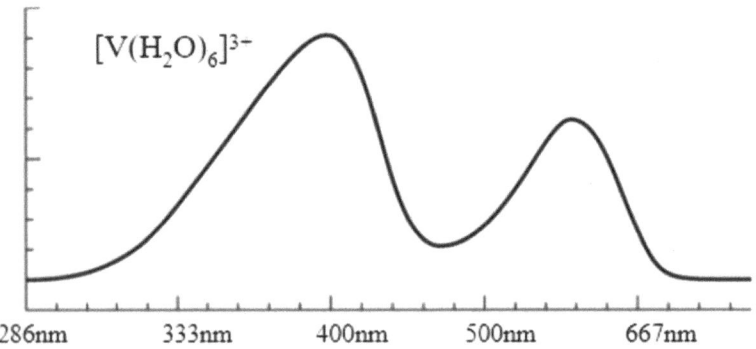

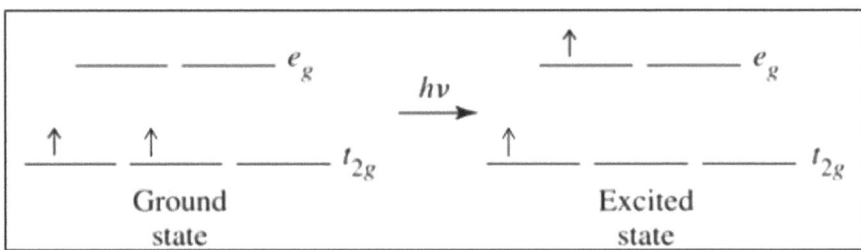

The explanation lies in the energetic difference if one electron is for example in d_{xy} and the excited one in d_{x2-y2} or in d_{z2}.

[12] *Which configuration would have a higher energy:*

$d_{xy} + d_{z2}$ *or* $d_{xy} + d_{x2-y2}$

(remember that electrons are repulsing each other !)

The d-d splitting depends on a) the charge on the metal and b) the nature of the ligand. Compare some Δo values for Cr(3+) complexes:

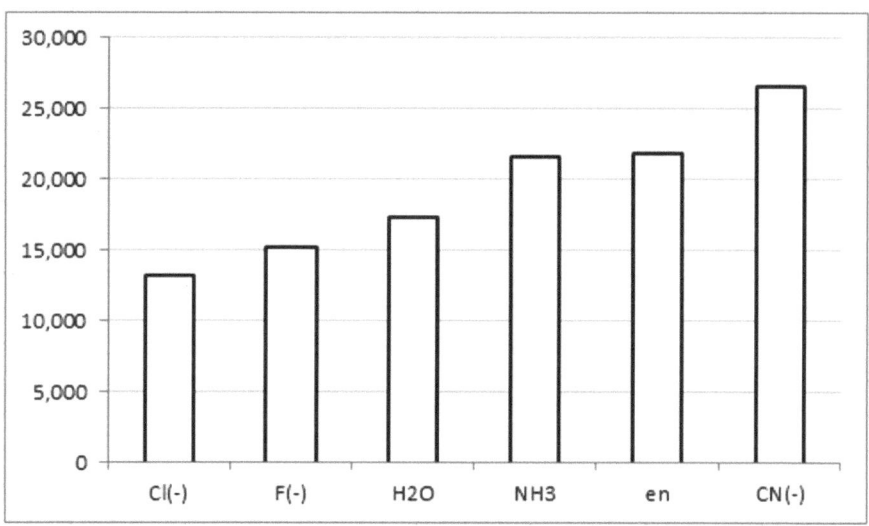

In general the more basic a ligand, the higher is the d-d splitting.

And the higher the charge on the metal – because in that case the ligands are pulled closer to the d-orbitals.

[13] *A sample of one Cr^{3+} complex is green and another is yellow. One is $[Cr(NH_3)_6]^{3+}$ and the other is $[CrF_6]^{3-}$. Which is which? [Determine the color of light absorbed to determine energy – a green complex absorbs red (less energetic), a yellow complex absorbs purple (more energetic)]*

3.2 TERM SYMBOLS FOR D ELECTRON CONFIGURATIONS

Because of these energetic differences between excited states, the electron configuration cannot just be written as $t_{2g}^1 e_g^1$. Instead a combination of symmetry symbols and spin multiplicity is used.

Electronic States for free ions

Instead to just write d^2 for an ion electron configuration, we can be more precise, following **Hund's rules** for the ground state:

- Electrons are aligned in such a way that they produce the maximum possible spin S = number of unpaired electrons * 1/2
- The angular momentum of all electrons (L) is the sum of all m_L quantum numbers
- The total angular momentum J = |L - S| for less than half filled orbitals and L + S for more than half filled

[14] *Calculate the total spin S, angular momentum L and term symbol for a d^2 ion*

If you followed the rules, then S = 1 and L = 3

Instead of S, the *multiplicity* is used to describe all possible combinations of the individual spins which is 2*S + 1.

And for L, symbols are used instead of numbers – which resemble the orbital names, but now capital letters are used.

L =	State
0	S
1	P
2	D
3	F

A d^2 ion in the ground state would then be described as ^3F

The multiplicity 1 is often called "singlet", 2 "doublet" and 3 "triplet"

Finally the total angular momentum J is the combination of L and S:

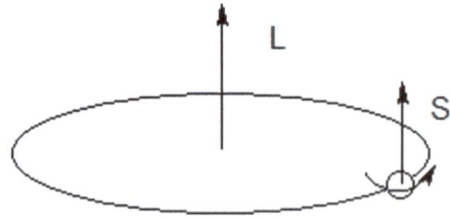

Both rotational movements of the electron create a combined momentum J:

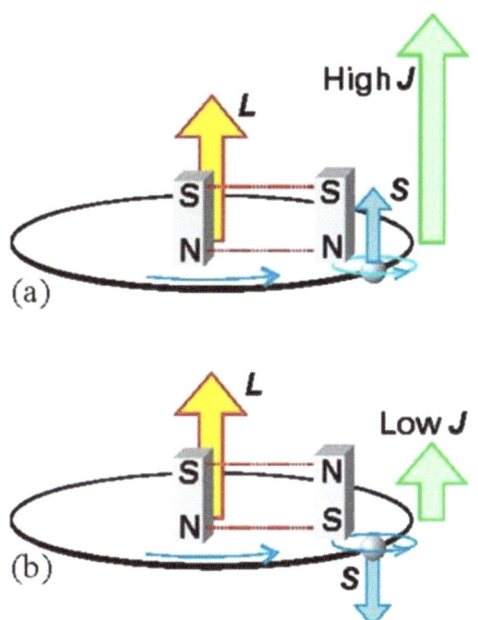

J can have the values: L + S, L + S -1, ... | L - S|

For d^2: J = 4, 3, 2

The orbitals are less than half filled, so the ground term is $^3F_2$

[15] *Determine the term symbols for the ground state of 2 free ions with d^4 and d^8*

Symmetry symbols for d-electrons in a crystal field

The "symmetry" of a configuration is simply evaluated by counting the possible different d-orbital occupations that have the same crystal field energy.

For the ground state of d^2 there would be three different possibilities for the two electrons in t_{2g}:

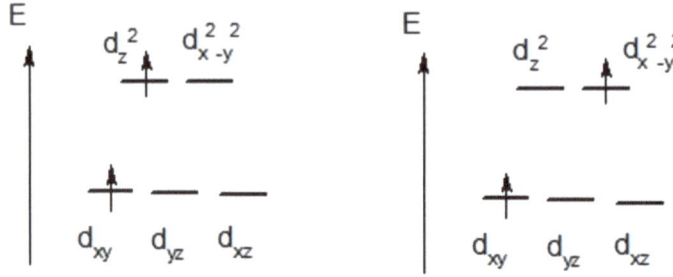

An energy state with 3 different possibilities is called "T", so the d^2 ground state could be describe as ^3T configuration

In the first case, the 2 electrons are in orbitals *perpendicular* to each other:

$(z^2 + xy)$ $(x^2-y^2 + xz)$ $(x^2-y^2 + yz)$

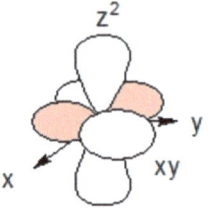

In another case, the 2 electrons are in orbitals in the *same plane*:

$(z^2 + xz)$ $(z^2 + yz)$ $(x^2-y^2 + xy)$

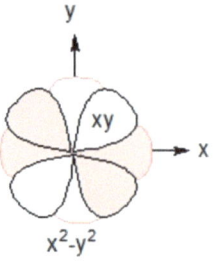

As we saw before, there is an energy difference between these two cases, which both can be described as ^3T: the second configuration has a higher energy due to the higher electron-electron repulsion.

Finally there can be a <u>second excited state</u> where both electrons are in the upper orbitals.

Since there is only one possibility to arrange the 2 electrons, the state can be described as ^3A.

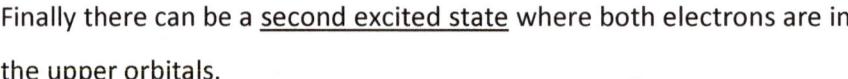

An electronic transition or excitation would be described not merely as

$t_{2g}^2 e_g^0 \rightarrow t_{2g}^1 e_g^1$

but more precisely as $^3T \rightarrow {}^3T$.

Now we can have 3 electron transitions:

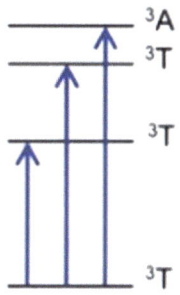

[16] *name the following electronic configurations:*

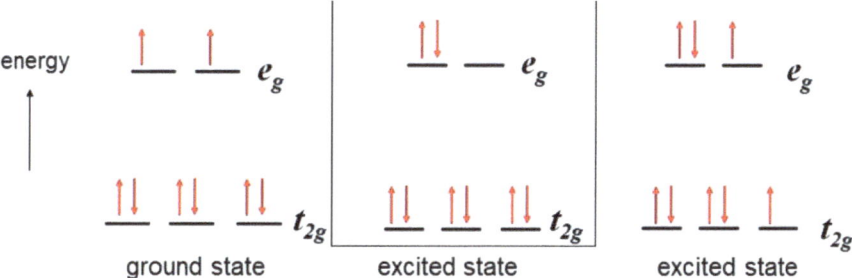

[17] *Draw the ground and excited states of a Ni^{2+} complex*

3.3 TANABE-SUGANO DIAGRAMS

In these diagrams we see the splitting of atomic terms, in this example of d^2, into different electron configurations under the influence of a ligand field.

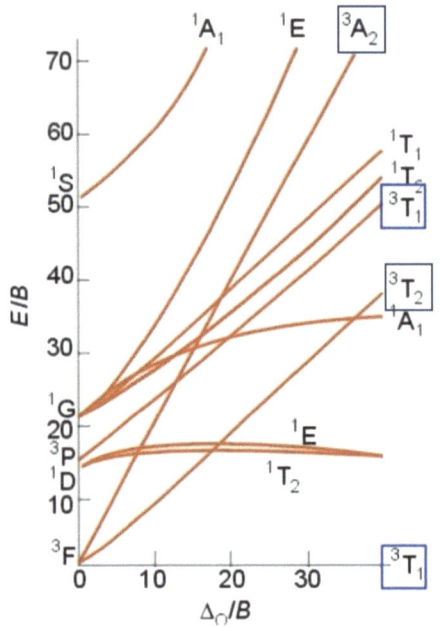

The x-axis starts with the atomic term of ^3F for a non-split d^2 configuration. When the outer ligand field increases, three excited states evolve, called $^3T_2$, $^3T_1$ and $^3A_2$.

An important rule for electronic transitions is the conservation of electron spin: it is actually "forbidden" for an electron to change its spin during a transition. This rule is quite strict, though small signals may be seen for this case.

Main transitions therefore can only take place between the ground state ^3T to the three excited states $^3T_2$, $^3T_1$ and $^3A_2$.

(Since $^3A_2$ is at high energy, this transition is most likely in the UV region of the spectrum)

<u>Use the Tanabe-Sugano diagram to find Δo</u>

In a real VIS spectrum of a d^2 V^{3+} complex we find 2 main peaks:

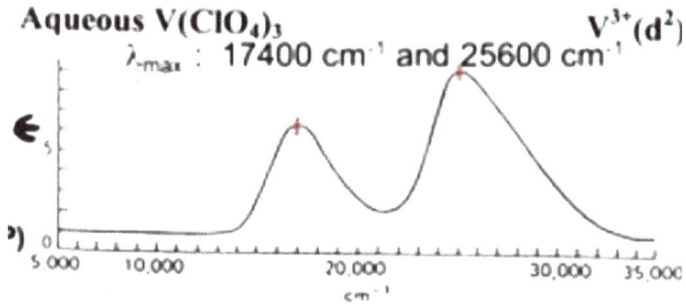

Aqueous $V(ClO_4)_3$ $V^{3+}(d^2)$
λ_{max} : 17400 cm^{-1} and 25600 cm^{-1}

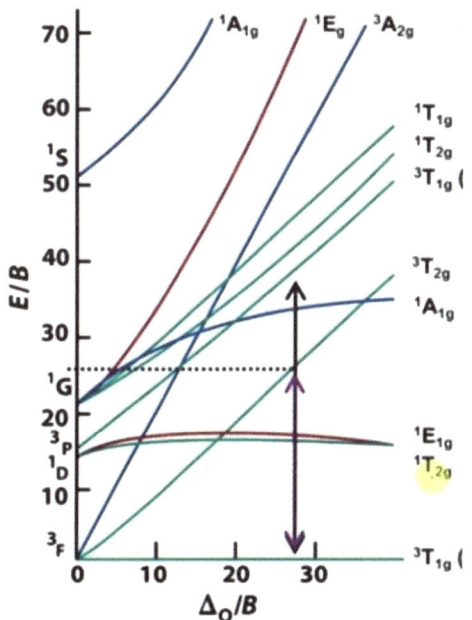

Step 1: find the ratio of the energies of the peaks:

25600/17400 = 1.47

Step 2: Use a ruler to find the same ratio for transitions in the diagram:

This happens at an x-value of:

$\Delta o/B = 27$

Step 3: Check the corresponding y-value E/B = 25.5 which belongs to the lower energy transition of 17'400 cm^{-1}

Step 4: combine both results:

E/B = 25.5 and $\Delta o/B = 27$

with E = 17'400 cm-1

=> B = 680 cm-1

With this B we can find Δo = B * 27 = 680 cm^{-1} * 27 = **18'360 cm^{-1}**

[18] *Use the same method to find Δo for a d^8 complex:*

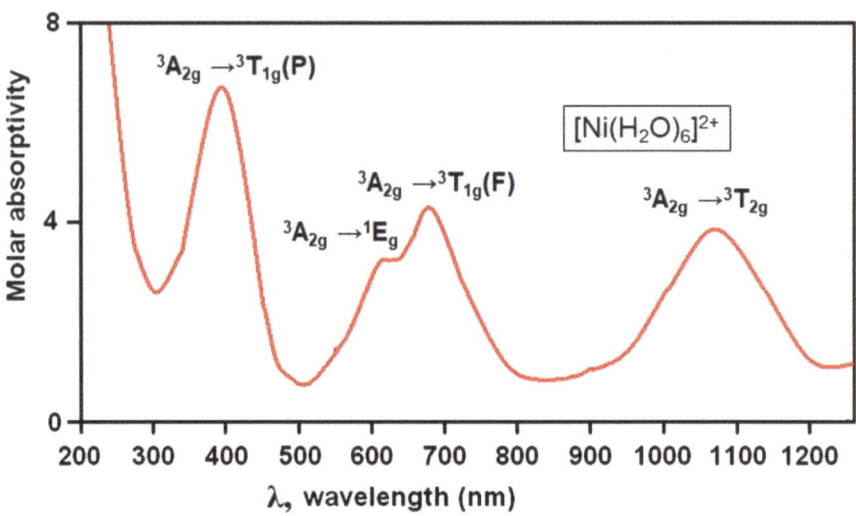

(The indication "P" and "F" comes from the origin of the atomic configuration and is beyond the scope of this tutorial – just remember that there are two different $^3T_{1g}$ states)

(http://wwwchem.uwimona.edu.jm/courses/ Tanabe-Sugano/TSHelp.html)

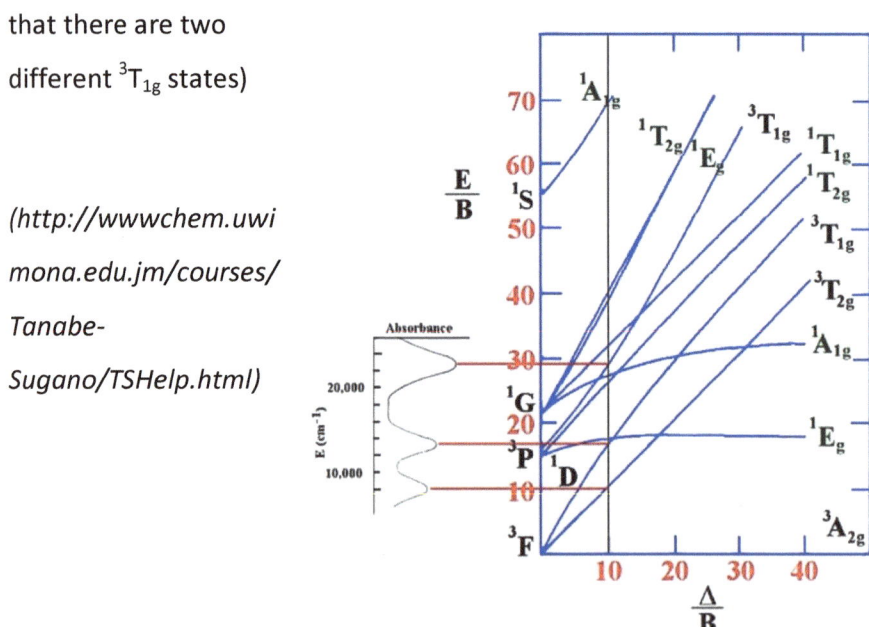

3.4 Selection rules

The electron transitions follow three rules:

1. The change in angular momentum Δl should be +/- 1

   That means:

   d -> d are <u>forbidden</u>

   p-> d and s-> p are allowed.

   Therefore ALL d-d transitions are forbidden according to this rule. That we detect a peak is due to the vibrational deformation that the orbitals undergo.

2. For molecules with an inversion center (as octahedral complexes, but not tetrahedral !) the <u>Laporte selection rule</u> is valid which states that transitions between states of the same inversion symmetry are forbidden:

   that means **g -> g** and u -> u are <u>forbidden</u>,

   but g -> u and u-> g are allowed.

   A transition between t_{2g} and e_g in octahedral complexes is therefore also always forbidden !

3. The strictest rule is the <u>spin selection rule</u>: $\Delta S = 0$

   Transitions between states with different spin multiplicities are forbidden.

   For example the transition $^3T_{1g} \rightarrow {}^1A_{1g}$ is forbidden,

   but $^3T_{1g} \rightarrow {}^3A_{2g}$ is allowed

Because all d-d transitions in octahedral complexes are forbidden according rule 1 and 2, the intensities are relatively small. Compared to that, d-d transitions in tetrahedral complexes violate only the first rule and are therefore of much higher intensity.

By vibrations, all molecules temporarily change their symmetry and make "forbidden" rules possible.

Example: d^3 complex:

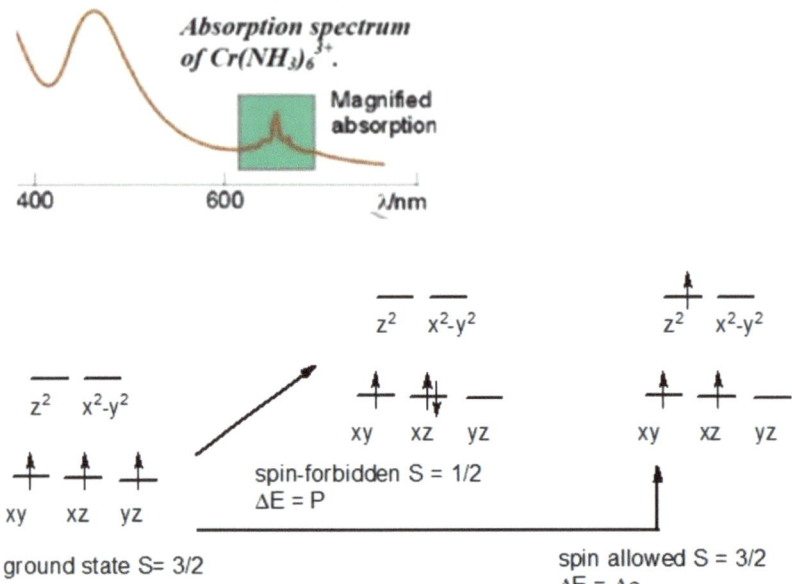

Compared to d-d transitions, there can be *"charge-transfer"* transitions between electrons on metal-orbitals and empty ligand orbitals or vice versa – these transitions are very intense since they do not violate the selection rules. Fortunately they lie at much higher energy and therefore normally in the UV-range of a spectrum and do not interfere with our d-d transitions.

A typical application that takes advantage of the different intensities of tetrahedral and octahedral complexes is a silica-gel with tetrachloride-cobaltate(II) as indicator for water:

$[CoCl_4]^{2-} + H_2O \longrightarrow [Co(H_2O)_6]^{2-}$

[17] *explain which color change we can expect when the indicator becomes wet. [take into account that Cl^- is a weaker ligand as H_2O]*

4 CRYSTAL FIELD STABILISATION ENERGY CFSE

Due to the d-electron splitting, the total energy of the d-electrons is often lowered. We can calculate this energy gain quite easily by adding up the energies of each d-electron.

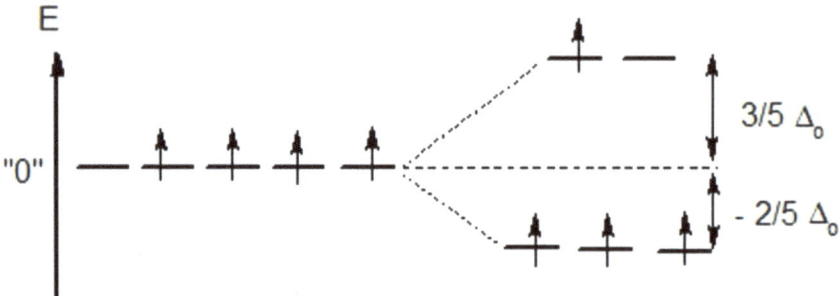

The splitting is symmetric around the original energy of the 5 d-orbitals. The 3 electrons in the lower level have the energy of

$3 * (-2/5\ \Delta o) = -6/5\ \Delta o$

and the one electron in the upper level is $+ 3/5\ \Delta o$.

The total CFSE therefore is $-6/5\ \Delta o + 3/5\ \Delta o =$ **$- 3/5\ \Delta o$**

The value of Δo depends on the nature of the metal and ligands.

In the case of <u>paired electrons</u>, we get a complication:

to pair 2 electrons, there is an additional energy necessary, the so-called *electron-pairing energy P*.

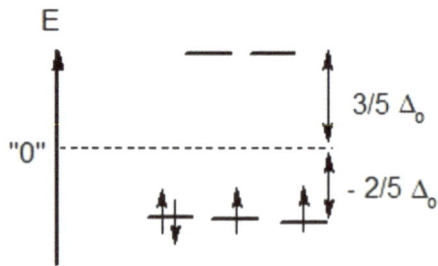

The CFSE in this example would be: $4 * (-2/5\ \Delta o) + P =$ **$-8/5\ \Delta o + P$**

P is normally in the magnitude of Δo. Here some examples:

Complex	Config.	Δo in cm^{-1}	P in cm^{-1}	spin
$[Fe(H_2O)_6]^{2+}$	d^6	10'400	17'600	High-spin
$[Fe(CN)_6]^{4-}$	d^6	32'850	17'600	Low-spin
$[CoF_6]^{3-}$	d^7	13'000	21'000	High-spin
$[Co(NH_3)_6]^{3-}$	d^7	23'000	21'000	Low-spin

[20] *Calculate the CFSE of $[Fe(H_2O)_6]^{2+}$ and $[Fe(CN)_6]^{4-}$ in cm^{-1}*

We see that in both cases the CFSE is negative – that means that the d-d splitting is energetically favored for these d^6 complexes.

For tetrahedral complexes, the Δt values are much smaller, for example:

M(H$_2$O)$_4$ complex	Δ_T in cm^{-1}
V^{2+}	5200
V^{3+}	8400
Cr^{2+}	6200

Cr^{3+}	7800
Fe^{2+}	4400
Fe^{3+}	6200

We see that the Δ_t is smaller than P, therefore **all tetrahedral complexes are high-spin** !

The splitting values Δ_T for the same cation are 4/9 of the Δ_O : (upper curve for octahedral complexes)

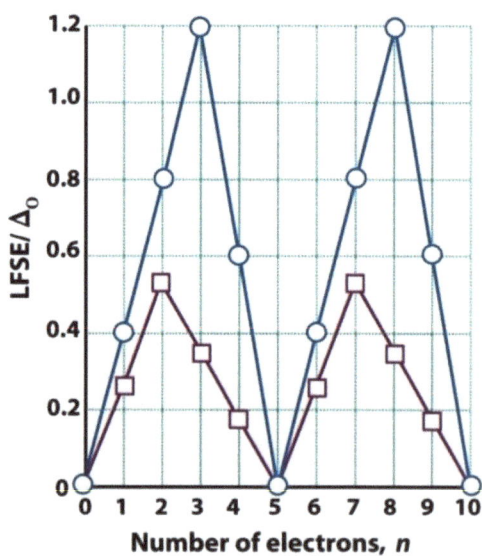

4.1 LATTICE ENERGY

The lattice energy is the energy that is needed to separate all ions in a crystal lattice into a gaseous state, like for example:

$CaCl_2$ (s) ---> Ca^{2+}(g) + 2 Cl^-(g)

This energy represents all electrostatic attractions in one mole of crystals – similar to the bond energy in one mole organic molecules.

For our daily experience, the *solution enthalpy* ΔH_{sol} is more common. It is related to the lattice energy by:

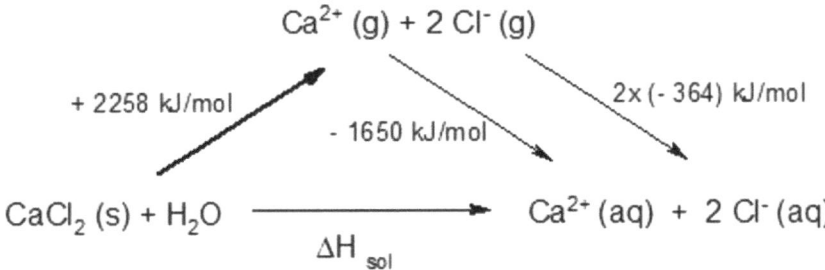

[21] *Calculate the solution enthalpy from the lattice and solvation energies of the ions – will the solution reaction of calcium chloride be exo- or endothermic ?*
(http://www.chemguide.co.uk/physical/energetics/solution.html)

This lattice energy in a series of different cations with the same anion like F^- should <u>increase</u> in the periodic table from left to right – because the size of the metal cations decreases.

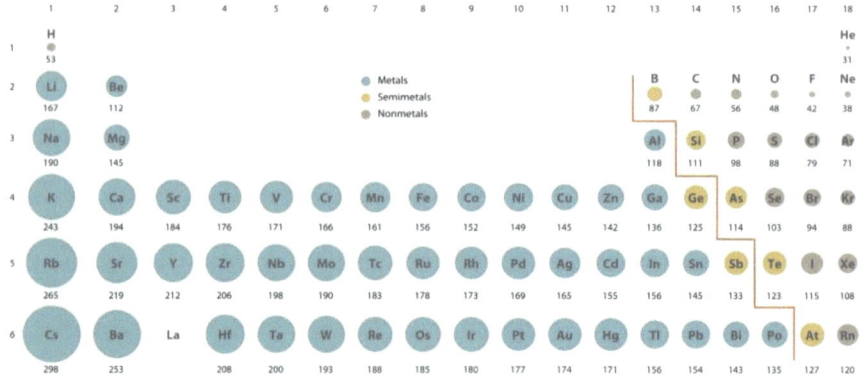

(https://chem.libretexts.org/LibreTexts/Howard_University/General_Chemistry%3A_An_Atoms_First_Approach/Unit_1)

Smaller cation size means that the ions in a lattice come in closer contact, therefore their attraction should be higher:

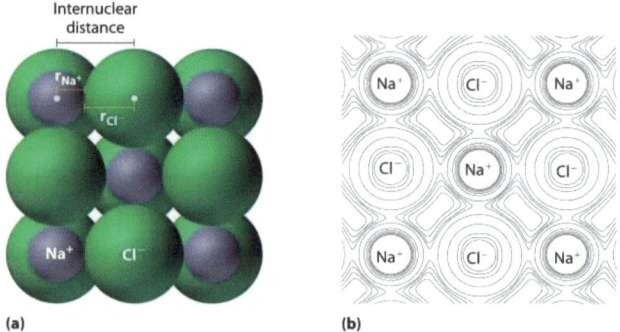

When we compare the lattice energies of MF_2 compounds (weak field ligand), then only Ca^{2+} (d0), Mn^{2+} (d5) an Zn^{2+} (d10) show the expected linear trend with decreasing ion radius. For all other configurations, there is a CFSE which has to be overcome in addition to the electrostatic attraction between the ions.

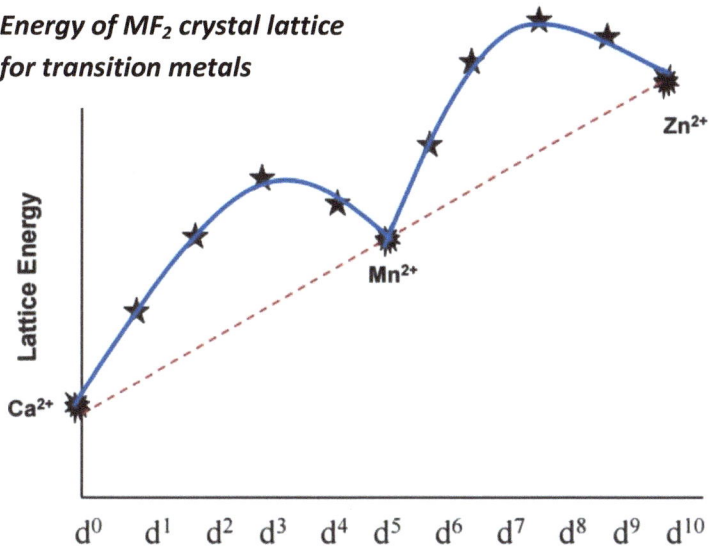

[22] *Explain why the lattice energy for d^3 and d^8 deviate the most from the expected trend.*

4.2 HYDRATION ENERGIES

Related to lattice energies are the energies that a metal cation gains when it is transferred from gaseous state to dissolved state in water:

M^{2+} (g) + H_2O (l) ---> $[M(H_2O)_6]^{2+}$ (aq)

This process is exothermic and should increase linear with decreasing ion size (smaller ions have shorter metal-water bonds). Again we can find a trend that deviates from the linear relation because of additional CFSE. Water as weak ligand causes high-spin complexes to form, therefore the Mn^{2+} d^5 coordination has no CFSE.

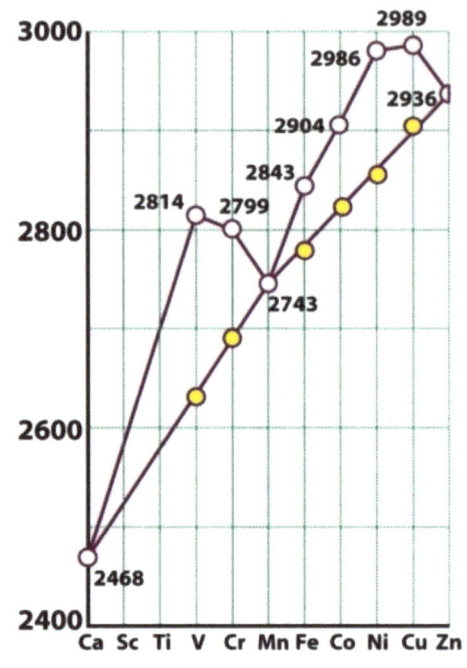

The greatest gain in energy experiences the V^{2+} (d^3) ion and Cu^{2+} (d^9)

[23] *explain arguing with CFSE – explain also why Cu^{2+} has an even higher stabilization than Ni^{2+}*

4.3 IONIC RADII OF CATIONS IN A CRYSTAL LATTICE

The CFSE also affects the ionic radii of metal cations itself in a lattice – imagine a chloride $M^{2+}Cl^{2-}$ lattice: the distance between the metal and the chloride should decrease across the period table due to the smaller metal cations. The measured values show again a double-curve, where metals with high CFSE move closer to the cations to increase this additional effect:

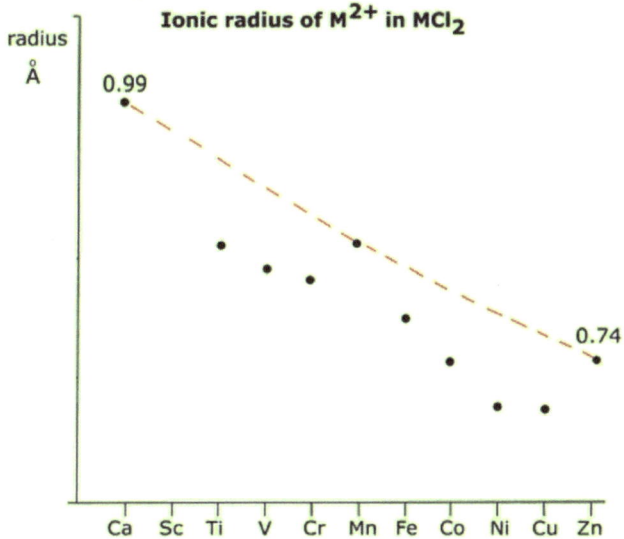

(http://www.mjmorris.staff.shef.ac.uk/teaching/CHM1002/lect3a.html)

Again the strongest deviations are found for d^3 and d^8 / d^9 compounds, which profit the most from a high splitting Δo.

[24] *Rationalize why V^{2+} has a lower ionic radius than Mn^{2+}, even though the electrostatic attraction of the valence electrons is lower.*

4.4 Spinel crystals

"Spinels" are lattices with the general formula A B_2 O_4 (A,B transition metals) – these solids are quite abundant in nature and are known generally as "gemstones".

When we consider a closest packing of oxygen ions, where the holes are occupied with metal cations (which are much smaller in size). In this kind of packing we find two different kind of holes – tetrahedral and octahedral:

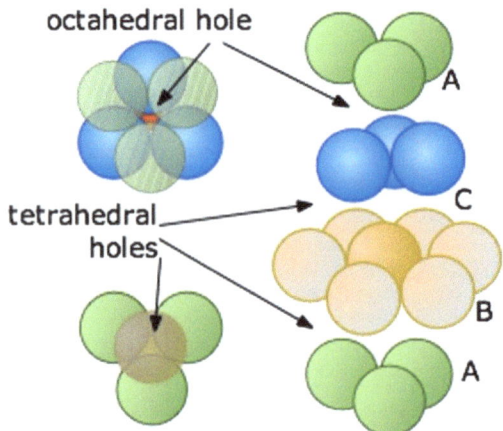

(https://chem.libretexts.org/Textbook_Maps/General_Chemistry_Textbook_Maps/Map%3A_Chem1_(Lower)/07%3A_Solids_and_Liquids/7.08%3A_Cubic_Lattices_and_Close_Packing)

We find that there are as **double as many octahedral than tetrahedral holes**.

In a spinel –type crystal it seems logic that one cation type sits in tetrahedral and another type in octahedral holes according to

$$A^{(2+)} B^{(3+)}{}_2 O_4$$

with $A^{(2+)}$ in tetrahedral and $B^{(3+)}$ in octahedral holes

But another configuration, known as "inverse spinel" can also be found where the +2 cations occupy the octahedral holes:

$$B^{(3+)}[A^{(2+)}B^{(3+)}] O_4$$

Which spinel type is formed, depends on the relative sizes of the ions A and B (the smaller cation prefers the tetrahedral sites) and – to a more extent - on the lowest crystal field stabilization energy (CFSE).

Take for example Mn_3O_4, which consists of two types of manganese ions:

$$Mn^{(2+)} Mn^{(3+)}{}_2 O_4 \text{ - OR: } Mn^{(3+)}[Mn^{(2+)} Mn^{(3+)}] O_4$$

[25] *Which one of the two molecules would have lower CFSE ?*

Quite surprisingly, the very simplified crystal field theory gives similar results to much more advanced chemical calculation methods, as shown in the article below:

J. Am. Chem. Soc. **1982**, *104*, 92–95

Role of the Crystal-Field Theory in Determining the Structures of Spinels

Jeremy K. Burdett,*[1a,b] Geoffrey D. Price,[1c] and Sarah L. Price[1b]

Contribution from the Departments of Chemistry and Geophysical Sciences, The University of Chicago, Chicago, Illinois 60637. Received April 8, 1981

Abstract: Pseudo-potential orbital radii r_s, r_p are used to construct an index $r_\sigma = r_s + r_p$. A plot of r_σ^A vs. r_σ^B for 172 chalconide spinels (AB_2X_4) leads to two well-defined areas which contain only normal or inverse spinels, with only four errors. At the boundary of the two regions the observed structures are generally in agreement with crystal-field ideas. The gross sorting is achieved without recourse either to the number of d electrons or an orbital radius r_d, which implies that it is the A,B s- and p-orbital energies which primarily determine coordination numbers in these systems. In fact, whereas d-orbital-based crystal-field ideas are only practically applicable to 74 transition metal containing examples of our data base (of which only 61, or 82%, are predicted correctly as normal or inverse variants), good structural sorting is achieved for all examples using r_σ plots (a 98% success rate). The relative (minor) importance of d orbitals and (major) importance of higher energy s, p orbitals on A,B is thus in accord with the relative energetic importance of these orbitals in ligand coordination. For the first time the reasons determining site preferences of non-transition-metal ions are identified.

[26] *Estimate the spinel structures for:*

(a) Fe_3O_4

(b) $CoFe_2O_4$

(c) $NiFe_2O_4$

(d) $NiMn_2O_4$

(e) $FeCr_2O_4$

(f) Co_3O_4

(http://www.adichemistry.com/inorganic/cochem/spinels/spinel-structures.html)

5 MAGNETIC PROPERTIES

CFT does not only tell us a lot about electronic and physical properties of transition metal compounds but does also help us to estimate magnetic properties.

We have to distinguish between three types of magnetism:

- ferromagnetism
- para- and dia-magnetism

In this booklet we consider only para- and dia-magnetism: they mean that a substance is attracted / repelled by an external magnetic pole, without being magnetic by itself.

5.1 MEASUREMENT OF MAGNETIC MOMENTS

The more unpaired electrons, the stronger is the paramagnetic effect (when a molecule has only paired electrons, it is repelled slightly by an external magnet – that effect is called dia-magnetism).

This attraction or repulsion of a substance can be measured by a <u>Gouy balance</u>:

http://wwwchem.uwimona.edu.jm/utils/gouy.html

5.2 PARAMAGNETISM OF TRANSITION METAL COMPLEXES

This kind of magnetism is based on <u>unpaired electrons</u> in the molecules: because of the electron spin, an unpaired electron has a magnetic moment μ_{spin}. In addition, the rotation of an electron around the nucleus also creates a second magnetic moment $\mu_{orbital}$:

For first-row transition metals, the spin magnetic moment is dominant – whereas for second and third row metals, the orbital moment becomes significant.

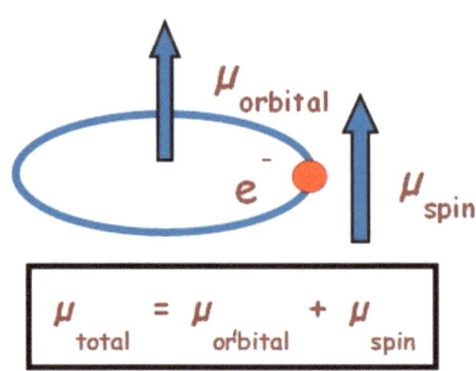

There is a simple estimation for the spin magnetic moment, the so-called "spin-only formula": $\mu = \sqrt{n*(n+2)}$

(n : number of unpaired electrons)

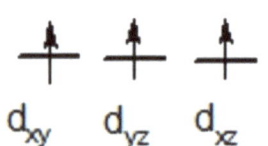

For example for a high spin Fe^{3+} ion in an octahedral ligand field:

Fe is in column 8 of the periodic table, so it has in total 8 valence electrons (s and d together) – therefore Fe^{3+} has 8 -3 = 5 valence electrons. High spin means that these electrons are evenly distributed over the d-electron levels.

We get 5 unpaired electrons, so according to the spin-only formula is:

$\mu = \sqrt{(5*7)} = 5.92$

The experimental value is 5.9, so we get a very good agreement in this case.

[27] *Find the electron configuration of the following complexes from the measured magnetic moment μ:*

- $[Cr(NH_3)_6]Cl_2$ $\mu = 4.85$
- $[V(NH_3)_6]Cl_2$ $\mu = 3.9$
- Co(II) complex $\mu = 4.0$
- $[Mn(NCS)_6]^{4-}$ $\mu = 6.06$

5.3 RUSSEL-SAUNDERS (LS) COUPLING

For one unpaired electron, there are two possibilities for the orientation of the magnetic moments coming from the electron spin and the orbital spin: the total magnetic moment J is the vector sum of these two moments.

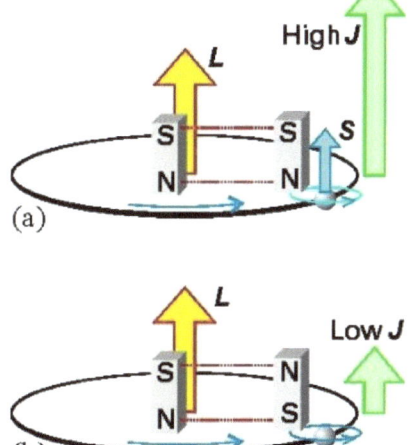

The orbital moment becomes strong for:

- second- and third-row transition metals
- low-spin d^5 and high-spin d^6 and d^7 complexes

How about the second point ?

The d_{x2-y2} orbital is special insofar that it extensively overlaps with the d_{xy} :

In the consequence we can imagine that a single electron in d_{xy} can use the d_{x2-y2} (empty or with one electron) to instigate a rotation around the nucleus.

A similar situation occurs when we have a high-spin d^6 or d^7 complex:

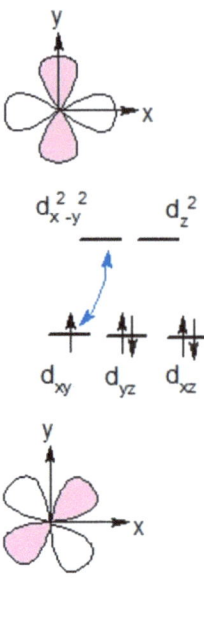

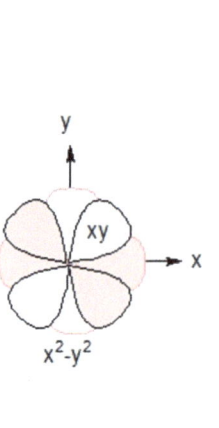

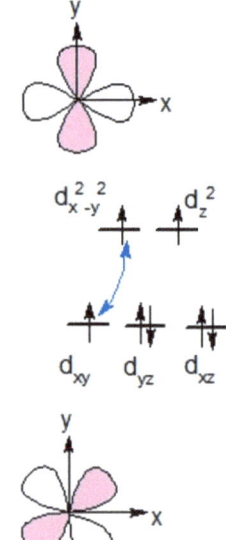

We can estimate the magnetic moment in the case of a strong LS coupling:

$$\mu = \mu_{spin}(1-\alpha\lambda/\Delta_o)$$

α: 4 for an A ground state and 2 for an E ground state

λ : spin-coupling coefficient in cm^{-1} that we have to look up

Ion	Ti^{3+}	V^{3+}	Cr^{3+}	Mn^{3+}	Fe^{2+}	Co^{2+}	Ni^{2+}	Cu^{2+}
d conf	d1	d2	d3	d4	d6	d7	d8	d9
λ (cm^{-1})	155	105	90	88	-102	-177	-315	-830

[28] *Compare the magnetic moment calculated by the spin-only formula and the corrected formula above for a Ni^{2+} complex. ($\Delta o = 11'500\ cm^{-1}$)*

6 ANSWERS

[1] Blue $CuSO_4 \cdot 5H_2O$ absorbs orange light which has a wavelength of about 600 nm. The energy of this absorbed light is:

$E = h \cdot c/\lambda = 4.14 \cdot 10^{-15}$ eV·s $\cdot\ 3 \cdot 10^8$ m/s $/\ 6 \cdot 10^{-7}$m $= 4.14 \cdot 3/6$ eV = **2.07 eV**

The wavenumber is equivalent to energy and used often in spectroscopy instead:

$\nu = 1/\lambda = 1 / 6 \cdot 10^{-7}$m $= 1/6 \cdot 10^{-5}$cm = **16'667 cm^{-1}**

[2] Chlorophyll b absorbs light at 500 nm, therefore the energy absorbed is:

$E = h \cdot c/\lambda = 4.14 \cdot 10^{-15}$ eV·s $\cdot\ 3 \cdot 10^8$ m/s $/\ 5 \cdot 10^{-7}$m = **2.48 eV**

[3] Contact between ligand orbitals and d_{z2} vs d_{xz}

Since the d_{xz} orbital has nodes at the points where ligands approach, its interaction is weak – but the d_{z2} orbital has its maximum where the ligands in z direction approach. Therefore the d_{z2} orbital is repulsed by the negative charge from the ligand and is raised in energy.

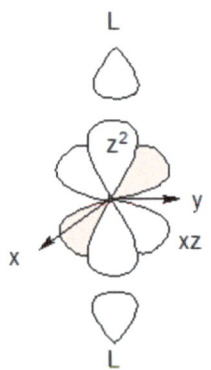

[4] The symmetry of the d_{x2-y2} and d_{z2} orbital in an octahedral molecule:

Character table for O_h point group

	E	8C$_3$	6C$_2$	6C$_4$	3C$_2$ =(C$_4$)^2	i	6S$_4$	8S$_6$	3σ$_h$	6σ$_d$	linear, rotations	quadratic
A$_{1g}$	1	1	1	1	1	1	1	1	1	1		x^2+y^2+z^2
A$_{2g}$	1	1	-1	-1	1	1	-1	1	1	-1		
E$_g$	2	-1	0	0	2	2	0	-1	2	0		(2z^2-x^2-y^2, x^2-y^2)
T$_{1g}$	3	0	-1	1	-1	3	1	0	-1	-1	(R$_x$, R$_y$, R$_z$)	
T$_{2g}$	3	0	1	-1	-1	3	-1	0	-1	1		(xz, yz, xy)
A$_{1u}$	1	1	1	1	1	-1	-1	-1	-1	-1		

xz, yz and xy appear as T$_{2g}$. x^2-y^2 as E$_g$.

z^2 appears under A$_{1g}$ and under E$_g$ – because we are looking for the combinations of z^2 and x^2-y^2, only the E$_g$ part is counting in our case.

[5] Δo for Ti^{3+} can be calculated directly from the UV/VIS spectrum because there is no d-d electron interaction in this case (d^1). The absorption maximum lies at about 500 nm, so the energy (as wavenumber) absorbed is:

1/λ = 1/5*10-7 m = 1/5*10-5 cm = 20'000 cm^{-1}

[6] Interaction between ligands in tetrahedral symmetry and d_{z2} vs d_{xz}

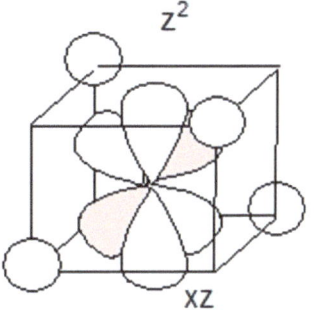

In contrast to the octahedral symmetry, it is harder to see which orbital comes closer to the ligands – in this case the **xz** is in closer proximity than the z^2, but the difference is smaller than in the octahedral case.

[7] From the picture in question [6] we can conclude that the difference in repulsion between the ligands and d_{z2} and d_{xz} is quite small. Therefore the Δ_t (energy difference between these orbitals) is also small compared to the octahedral geometry.

With a small energy gap between the d-orbitals, electrons will distribute evenly and avoid a pairing. So tetrahedral complexes are always high-spin.

[8] In a square-planar configuration, 8 electrons are clearly the preferred configuration – whereas in a tetrahedral case, any electron count is possible since the energy gap Δ_t is quite small:

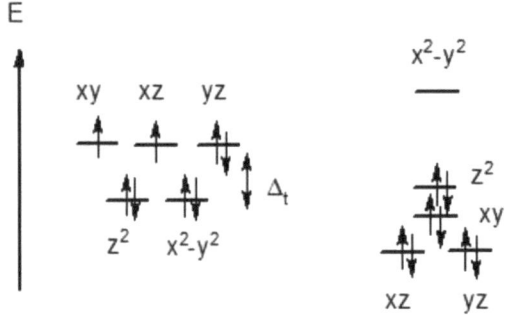

[9] Cu(II) complexes gain extra stability by the Jahn-Teller effect:

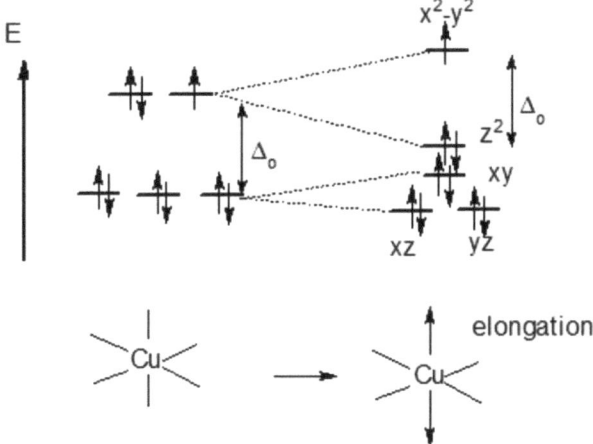

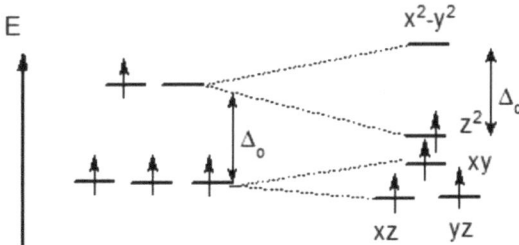

This effect does not produce extra energy for a d^8 or d^{10} configuration as in Ni(II) and Zn(II) – in Ni(II) the 2 "upper" electrons would be in d_{z2} and d_{x2-y2} (they would not pair up)

[10] When we look at the energy diagrams of a high-spin d^4 complex:

we can detect that a deformation lowers the energy of the single electron in d_{z2} considerably. But if this electron is in the lower xy, xz or yz d-orbital, no energy gain can be observed by a Jahn-Teller distortion.

[11] Increasing energy of light means that the color will change to shorter wavelengths, so there is a **blue**-shift.

[12] Consider that the d-orbitals live in different planes in space – the xy orbital is perpendicular to the z^2, but it is in the same plane as the x^2-y^2. Therefore the combination of xy and x^2-y^2 is at higher energy:

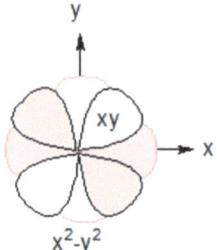

[13] A complex that looks green, absorbs red color, which is low-energetic, whereas a yellow one absorbs purple with much higher energy. That means that the green looking complex has a lower Δ_o than the other. Therefore the ligand-field is weaker. From the spectroscopic series we should remember that the halogens cause the smallest d-d splitting, so the green complex should be $[CrF_6]^{3-}$. The ammonium ligand in $[Cr(NH_3)_6]^{3+}$ causes much higher splitting which results in a yellow color.

[14] The electrons in a d^2 atom/ion are distributed according to Hund's rules:

The total spin is $S = \frac{1}{2} + \frac{1}{2} = 1$ and the multiplicity $2S + 1 = 3$.

The orbital momentum $L = 2 + 1 = 3$.

The combination $J = L + S = 4, 3, 2, 1, 0$. Since the shell is less than half filled, the ground state has $J = 4$ (highest J value)

So the term symbol is $^3F_4$

[15] Term symbols for d^4 and d^8 ion:

<u>d^4 case:</u>

$S = 4 * \frac{1}{2} = 2$

Multiplicity: $2S + 1 = 5$

$L = 2 + 1 + 0 - 1 = 2 \Rightarrow D$

$J = L+S, L+S-1, L+S-2, \ldots 0 = 4, 3, 2, 1, 0$

Term symbol: $^5D_4$

(ground state with highest J because more than half filled)

<u>d^8 case:</u> $S = 2 * \frac{1}{2} = 1 \Rightarrow$ Multiplicity: $2S + 1 = 3$

$L = 2*2 + 2*1 + 2*0 -1 -2 = 3 \Rightarrow F$

$J = L + S \ldots L - S = 4, 3, 2$

The shell is more than half-filled, so the ground state has highest J: $^3F_4$

[16] <u>Case 1</u>: $S = 2 * \frac{1}{2} = 1 \Rightarrow$ multiplicity $2S + 1 = 3$

Only 1 possibility to distribute the electrons $\Rightarrow {}^3A$

<u>Case 2</u>: $S = 0 \Rightarrow$ multiplicity $2S + 1 = 1$

2 possibilities to draw the configuration (either z^2 double occupied or x^2-y^2) $\Rightarrow {}^1E$

<u>Case 3</u>: $S = 2 * \frac{1}{2} = 1 \Rightarrow$ multiplicity $2S + 1 = 3$

3 possibilities to draw the configuration (leaving the electrons in eg constant, so actually 2 * 3 possibilities exist, but these don't have the same energy !) $\Rightarrow {}^3T$

[17] The ground state has spin 1 and multiplicity 2S + 1 = 3

In the ground state, only one configuration is possible, therefore the term symbol is A:

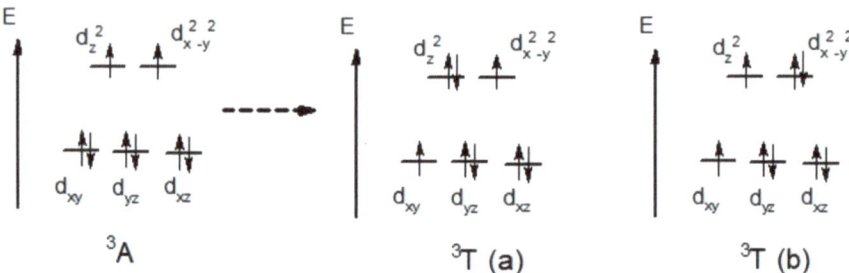

Spin allowed transitions (total spin remains 1) are possible between the ground state 3A and two different excited states 3T.

These two excited stated do not have the same energy because the single electron In eg interacts differently with, for example, the one in the d_{xy} orbital: if the single electron is in d_{z2}, then it is in a different plane, but if it is in d_{x2-y2} it is in the same plane and causes more repulsion

[18] Use the Tanabe-Sugano diagram for d^8 complexes:

Consider the peaks in UV and VIS at about 400 nm and 680 nm.

The corresponding energies (wavenumbers) are

$1/4*10^{-5}$ cm = 25'000 cm^{-1} and $1/6.8*10^{-5}$ cm = 14'700 cm^{-1}.

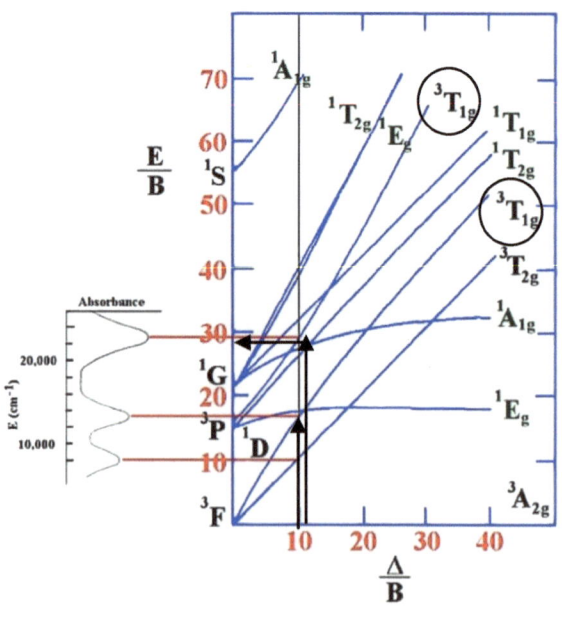

The energy ratio is 25'000/14'700 = 1.7

This ratio can be found at: $\Delta o/B = 10$

One value on the y-axis (E/B) is 29. With $E = 25'000$ cm^{-1} we get $B = 25'000$ cm$^{-1}/29 = 862$ cm^{-1} and therefore **Δo = B * 10 = 8620 cm^{-1}**

(Alternatively we could have used another of the 2 possible peak combinations)

[19] The Co(2+) ion has 7 valence electrons (Co is in column 9 in the periodic table) – the tetrahedral complex has a low energy gap Δt because of a) Cl- is a ligand that causes low splitting and b) tetrahedral complexes in general have a low d-d splitting. In addition the d-d transition is not Laporte forbidden. Therefore we can expect a strong color in the blue region (red color should be absorbed). Adding water changes the complex to a Co(2+) octahedral complex, where the d-d transition is forbidden and the energy gap is high.

[20] The water complex has a low splitting Δo and is therefore high-spin compared to the low spin cyano complex:

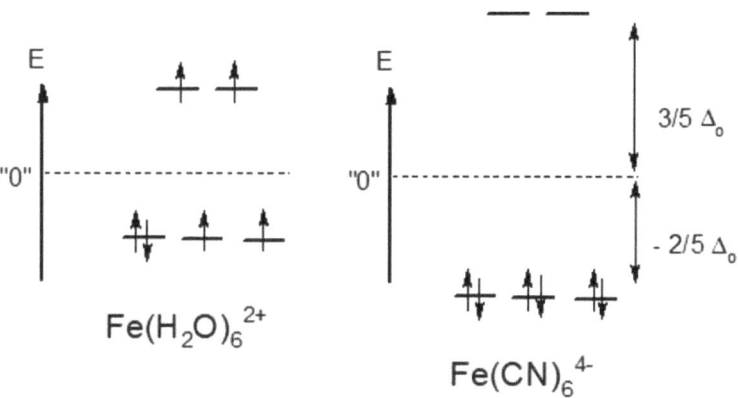

The crystal field stabilization energies are therefore:

(a) water complex:

$CFSE = 4 \times (-2/5\Delta o) + 2 \times (3/5\Delta o) + P =$

$= 4 \times (-2/5\ 10'400\ cm^{-1}) + 2 \times (3/5\ 10'400\ cm^{-1}) + 17600\ cm^{-1}$

$= +13'400\ cm^{-1}$

(b) cyano complex:

$CFSE = 6 \times (-2/5\Delta o) + 3 \times P$

$= 6 \times (-2/5\ 32'850\ cm^{-1}) + 3 \times 17'600\ cm^{-1}$

$= -26'040\ cm^{-1}$

Therefore the CFSE for the cyano complex is about 3 times higher.

[21] The solution enthalpy of $CaCl_2$ can be calculated from the lattice energy of the crystal (2258 kcal/mol) and the hydration energies of the individual ions (-1650 kcal/mol from Ca2+ and two times -364 kcal/mol for the two Cl-). The solution energy therefore can be calculated:

$\Delta H_{sol} = (2258 - 1650 - 2 \times 364)\ kcal/mol = -120\ kcal/mol$

Because the enthalpy is negative, the solution of $CaCl_2$ in water is an exothermic reaction.

[22] For high spin complexes, the highest CFSE we can get is at d^3. For more than 5 electrons, d^8 has the highest CFSE:

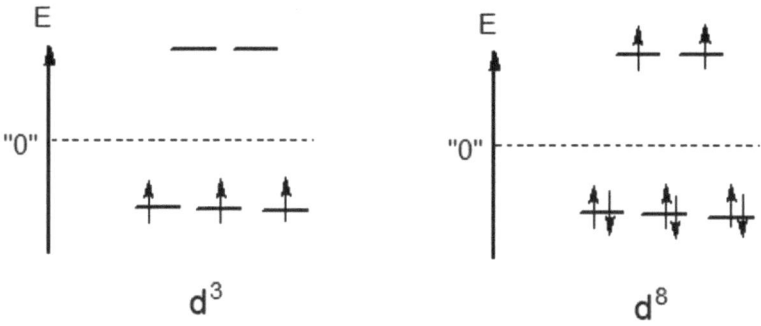

[23] The hydration energies of M^{2+} ions should rise continually when going from Sc^{2+} to Zn^{2+} because the ionic radii continually decrease (due to increasing effective nuclear charge) – then the distances between water molecules and the ions decrease. In an experiment we see that there is an additional energy that arises from the hydration of metal ions - and that additional energy is highest for d^3 and d^8/d^9 configuration. This can be explained by highest CFSE for these two configurations (see previous question). Cu^{2+} as d^9 ion gets a small extra stabilization due to the Jahn-Teller Effect – distortion of the octahedral geometry can lower the energy of one electron even further.

[24] The ionic radii of M^{2+} ions should decrease continually in the period Sc^{2+} to Zn^{2+}. This is because the effective nuclear charge of the ions continually increase. In reality we find that the ionic radii of MCl_2 compounds become even lower with a minimum for d^3 and d^8 configuration. We can explain this trend with a lower M-Cl distance for these configurations because in that case the Δo increases and therefore increases the crystal field stabilization.

[25] Compare the CFSE of two Mn^{3+} in octahedral position with one Mn^{2+} and one Mn^{3+} in octahedral position:

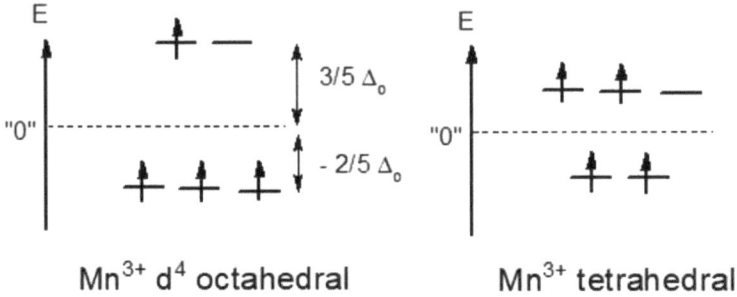

The CFSE for 2 atoms Mn^{3+} in octahedral position ("normal" spinel):

$2 \times [\, 3 \times (-2/5\, \Delta o) + 3/5\, \Delta o \,] =$ **-6/5 Δo**

Compared to one Mn^{3+} atom in octahedral position ("inverse" spinel):

CFSE(1) = $3 \times (-2/5\, \Delta o) + 3/5\, \Delta o =$ **-3/5 Δo**

and another in tetrahedral position:

CFSE(2) = $2 \times (-3/5\Delta t) + 2 \times (2/5\Delta t) =$ **- 2/5 Δt**

Since Mn^{2+} (d^5) has always no CFSE, the combination of two Mn^{3+} ions in octahedral position has higher CFSE.

Therefore the <u>normal spinel</u> Mn(II) [Mn(III)]$_2$ O$_4$ should be the natural form.

[26] Similar to the previous question we can calculate the CFSE for the octahedral ions (because $\Delta o > \Delta t$):

(a) Fe(II) Fe(III)$_2$ O$_4$ or Fe(III)[Fe(II) Fe(III)] O$_4$

CFSE for 2 Fe^{3+} in octahedral position = **0** (d^5)

and for Fe^{2+} and Fe^{3+} in octahedral (d^6) = 4 x (-2/5 Δo) + 2 x (3/5 Δo) = **-2/5 Δo (+ P)** Therefore the <u>inverse spinel structure</u> with Fe(II) in octahedral position has higher CFSE

(b) Co(II) Fe(III)$_2$ O$_4$ or Fe(III)[Co(II) Fe(III)] O$_4$

From (a) we learned that Fe(III) has no CFSE. Therefore the <u>inverse structure</u> with CFSE (d7) = 5 x (-2/5Δo) + 2 x (3/5Δo) = -4/5Δo (+ 2P) should be preferred.

(c) Ni(II) Fe(III)$_2$ O$_4$ or Fe(III) [Ni(II) Fe(III)] O$_4$

Since Fe(III) does not have a CFSE, here again the <u>inverse spinel</u> with Ni^{2+} in octahedral position should be preferred:

CFSE (Ni2+ = d8) = -6 x (2/5Δo) + 2 x (3/5Δo) = -6/5Δo (+ 3P)

(d) Ni(II) Mn(III)$_2$ O$_4$ or Mn(III) [Ni(II)Mn(III)] O$_4$

CFSE for 2 Mn^{3+} ions (d^4) = 2 x[3 x (-2/5Δo) + 3/5Δo] = **-6/5Δo**

and for one Mn3+ and one Ni2+: [3 x (-2/5Δo) + 3/5Δo] (d^4) + [6 x (-2/5Δo) + 2 x (3/5Δo)] (d^8) = **-9/5Δo (+ 3 P)**

In this case the CFSE is again higher for the <u>inverse spinel</u>.

(e) Fe(II) Cr(III)$_2$ O$_4$ or Cr(III) [Fe(II)Cr(III)] O$_4$

CFSE for 2 Cr^{3+} in octahedral configuration (d^3) = 2 x [3 x (-2/5Δo)] = **- 12/5Δo**

compared to Cr^{3+} and Fe^{2+} (d^3 and d^6) = [3 x (-2/5Δo)] +

$[4 \times (-2/5\Delta o) + 2 \times (3/5\Delta o)] = -8/5\Delta o\ (+ P)$

From this calculation we can conclude that this molecule should exist as a <u>normal spinel</u>.

(f) Co(II) Co(III)$_2$ O$_4$ or Co(III) [Co(II) Co(III)] O$_4$

CFSE for 2 Co^{3+} (d^6) = $2 \times [\ 4 \times (-2/5\Delta o) + 2 \times (3/5\Delta o)] = $ **-4/5 Δo (+ 2 P)**

compared to Co^{3+} (d^6) and Co^{2+} (d^7) = $[4 \times (-2/5\Delta o) + 2 \times (3/5\Delta o)\ ] + $
$[\ 5 \times (-2/5\Delta o) + 2 \times (3/5\Delta o)\ ] = $ **-6/5Δo (+ 3 P)**

Here the <u>inverse spinel</u> structure is more likely.

[27] (a) Cr^{2+} is d^4 => with 4 unpaired electrons: $\mu = \sqrt{4 \times 5} = 4.5$

=> high spin

(b) V^{2+} is d^3 => 3 unpaired electrons: $\mu = \sqrt{3 \times 4} = 3.5$

=> only high spin possible

(c) Co^{2+} is d^7 => with 3 unpaired electrons in high spin: $\mu = \sqrt{3 \times 4} = 3.5$

=> high spin configuration

(d) [Mn(NCS)$_6$]$^{4-}$ -> the ligands have 6 neg. charges,

then the Mn ion is Mn^{2+} which is d^5

If all electrons are unpaired (high-spin) then: $\mu = \sqrt{5 \times 6} = 5.5$

In a low-spin configuration there would be only 1 unpaired electron and the magnetic moment would be: $\mu = \sqrt{1 \times 2} = 1.4$

The experimental value of 6.06 indicates therefore a high-spin configuration.

[28] An octahedral Ni^{2+} complex has the ground state:

This complex has 2 unpaired electrons, so the spin-only formula gives us an estimation of

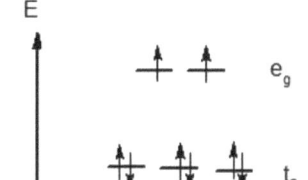

$\mu_{spin} = \sqrt{(2*4)} = \mathbf{2.83}$

For the LS coupling correction, we have to look up several values :

There is only one possibility to draw this configuration, so the ground state is an "A". Therefore $\alpha = 4$.

From the table on page..... we find that for Ni^{2+} the coefficient for the LS coupling is 315 cm^{-1}.

Knowing the octahedral splitting energy Δ_o = 11'500 cm^{-1} we can apply the estimation formula: $\mu = \mu_{spin}(1-\alpha\lambda/\Delta_o) =$

$= 2.83 * (1 - 4*315/11500) = \mathbf{3.14}$

The différence is not very big, but significant.

=============== CONGRATULATIONS ===============

You managed to go through the basics of Crystal Field Theory !
Now you are challenged to apply this knowledge to your day-to-day research

ABOUT THE AUTHOR

This booklet was written from a chemist for chemists.

I studied chemistry at the university of Constance, Germany and got my PhD from there in Inorganic and Theoretical Chemistry in 1988.
Later I worked as development chemist at Ciba-Geigy, Sandoz and Novartis in Switzerland for 10 years.
Afterwards I concentrated on teaching in various countries, right now at the University of Phayao at Thailand.

Christoph Sontag

www.ingramcontent.com/pod-product-compliance
Lightning Source LLC
Chambersburg PA
CBHW041107180526
45172CB00001B/144